Wings over Westray

Best Wishes

Jack Scott

Wings over Westray

By

Jack Scott

ISBN 0 9531563 0 3

Published by:
Information PLUS,
Finstown, Orkney, KW17 2LH

Printed by:
The Orcadian Limited, Hell's Half Acre,
Hatston, Kirkwall, Orkney, KW15 1DW

For Nan.

Cover Photograph Hotel India coming in to land at Westray, with Papa Westray in the distance. 1977.

Contents

Acknowledgements 6
Wings over Westray 7
Wings at war 8
Loganair 10
The flying doctor 13
Going solo 15
Getting wings 17
Calf over Calfsound 20
The Rosemary 22
Heavy weights 24
Gusty Grimsetter 26
Flying south for winter 29
The flying vet 32
Air shows 51
Friend or Foe in the Westray Firth 55
Surprise surprise! 56
The man fae the Ministry 58
Nuts, bolts and loud bangs 61
Retirement 62

Photographs 33 - 48

Acknowledgements

Writing this book is really a symptom of my whole life and involves only "doing as I am told". I should thank all the folk who have requested that I write these stories down, and those who have secretly recorded me telling stories to friends and visitors.

I am always grateful to Dr George Mears and Bob Learmonth who are no longer with us but between them got me flying. To Dr. Tim Wright who took over the annual medicals after Dr Mears left and continued as long as I flew.

I also want to thank all the folk who flew with me over the years and especially those who agreed to appear in this book. There are many stories yet untold but the way my memory works they are likely to remain so. Technical help from Alastair & Anne Cormack, Bob Tulloch and Colin Rendall was appreciated, as was the use of photographs by Dougie Shearer. I appreciate assistance from Orkney Islands Council, and Andy Alsop for his kind cover notes. As a Loganair pilot 30 years ago, he was one of my first flying friends and a great encouragement.

Final thanks to Fiona for typing up the manuscript from tapes and pestering me to "Haad gaan".

Jack Scott

Wings over Westray

I had always been interested in flying but for most of my life I never had the opportunity nor the finance to make a start.

Most of my early days were spent in Papa Stronsay, a very small island off Stronsay. I can recall time spent watching the sea gulls swoop and glide on the currents of air at the shore - I envied them their freedom, especially as my brothers and I were confined to the island while they could nip over to the village of Whitehall on Stronsay whenever they chose. Those were in the days of herring fishing however and I expect the gulls had all their requirements met on Papa Stronsay.

My first close up experience of an aircraft was when I was about 13. By this time I'd had to leave home to attend school at Stromness Academy. Although I missed home I was living with my uncle on the farm of Howe near Stromness. One evening a small plane came in and landed on one of the fields down near the Bush. We discovered it was Gandar Dower and Captain Starling who had flown up to look for a suitable landing strip for a proposed air service.

They wanted to secure the plane for the night and asked us if we could carry up stones from the beach to weigh the plane down. We all helped and I still remember Gandar Dower watching me struggling up the banks with heavy stones and telling me how strong I was for a schoolboy.

They did eventually start the Air Service which flew from Aberdeen to Stromness. It continued for a considerable number of years under the name of Allied Airways, flying De Havilland Dragon aircraft which were well suited to the grass runways. These planes were the same as the ones flown by Captain Fresson to Orkney.

The following year my father had bought the farm of Skaill on the island of Westray and my brothers were to attend school there. My father was not getting any younger and it was felt that I should leave the Academy and go to help him on the farm. I had thought of pursuing a career in medicine but for a young boy of 14 there was a certain appeal in leaving school to start work on the new farm. It also opened the way to my next meeting with an aircraft - and even later it had me meeting aircraft by appointment twice every day when an airstrip was established on our farm.

Wings at War

War broke out on my eighteenth birthday. I was never called up because of shortage of farm workers. The only plane to land with us during the war was a Swordfish which had been dive-bombing out past Noup. During the exercise the plane's engine had over heated and the pilot was forced to come down in our West Lands until it cooled off. When he felt it had cooled sufficiently he got some of us men to come and help him start it - it had a flywheel start and although we did get it going he decided it was damaged, having been almost seized.

Captain Curtis was the pilot and his colleague, a fellow by the name of Webb was the observor. We got to know both men and when 3 engineers were sent out with a replacement engine they seemed to enjoy the break. They enjoyed it so much that we wondered if the job would ever be done, and they took almost a week to get the new engine in.

Captain Curtis was flown out to reclaim his plane, and although he was not completely satisfied, he set off for Kirkwall and arrived safely. That was not the last we saw of him as he often flew over the farm and persisted in buzzing the house until somebody would wave to him. If all the men were away from the farm, one of the servant girls would be sent out to wave so that folk could get on with their work. The other airman kept up correspondence with my father for a long time and visited us at the end of the war. They must have had happy memories of Westray for the last time they wrote their house was called "Pierowall".

Captain Barclay came to Westray with a twin engine Dragon aircraft looking for a field to be used for an emergency airstrip for ambulance work. This was just after the last war and arose because of public concern over the issue. He landed on one of our fields which later became part of the airstrip which operates there at present. Captain

Barclay had done many ambulance flights in the Western Isles and was an expert in short field operations.

After landing, he invited my father and me to join him in the air for a circuit but my father wasn't interested. When he saw how disappointed I looked, however, he consented and that became my first flight which I greatly enjoyed.

Loganair

Much later in the sixties there was again interest in forming an air service in the North Isles. At this time the only transport to the Isles was the steamer sailing from Kirkwall and travelling round at least 5 islands before heading back to town. It followed that there was seldom a direct trip to town and no such thing as a day return apart from the County Show day and the annual "Trip Day" where each island got a turn of visiting town for the day.

I had become the Councillor for Westray in 1962, and spent many hours on the steamer trying to attend meetings. If the boat didn't suit I could easily spend 7 hours on the boat to get to a meeting with two overnight stops and 7 hours getting home again. I had also started to build a cabin cruiser in the garage and by 1967 I was ready to launch it. I could travel to Kirkwall in 2 hours in time for my meetings and also spend the night on board in Kirkwall if need be.

However by 1967 the venture between Orkney County Council and Loganair was almost at the point of take-off. There had been air fields identified in Stronsay, Sanday, Westray, Papa Westray, North Ronaldsay and Eday and an agreement reached with Orkney Islands Shipping Company for Loganair to operate the service on their behalf.

The field used by Captain Barclay was the one chosen for Westray and there was quite a bit of work done to get it ready for the start of the scheduled service. James Melrose the County Engineer was keen to fill one wet corner of the field with stone but I persuaded him to drain it instead and thankfully this worked.

Westray was a week late with the start of the service - the CAA decided at the last minute that one of the runways was too short so we had the county road men busy on the farm for a week moving a dry stone dyke and fence back several yards. It would have been easier to move the dyke before they had painted the red and white markings on the stones but somehow they managed and soon they pronounced the field ready.

Loganair then came out on a trial flight to check the strip. On the day that they flew out, our daughter Fiona who was 7 was quite ill with a chest infection. When the GP heard the plane was coming out that day he arranged for her to be admitted to the Balfour Hospital and she became one of the first to benefit from a speedy and comfortable admission to hospital by air ambulance. While she was an in-patient she had a visit by Willie Ross the Secretary of State for Scotland who was touring the hospital .

The air strip was provided with a wooden hut for the fire tender and for the reception of passengers. There were also two smaller huts with Elsan chemical toilets provided for the nervous traveller. When the fire tender arrived it was mounted on a trolley which we had to tow behind the Land Rover in the event of a fire. Unfortunately it was too wide to get in to the shed so I took it up to my workshop on the farm and took several inches off the trailer's width. I later heard that they'd had the same problem on all the other strips and had solved it by taking the facings off the shed doors! It was always a tricky business to reverse the tender back into the shed with a crowd watching.

There were, of course, training sessions for all of us so that we could operate the machinery, provide emergency first aid and also light flares to enable the pilot to land in darkness. Other duties included weighing all the luggage and passengers prior to them boarding the plane. All the holiday makers protested that their weight had gone up several pounds, blaming it on Westray hospitality.

I decided that such a new service was unlikely to have a major disaster on the first few trips and decided that I would attend my next few meetings by plane rather than boat. Initially the service would touch down on several strips on the way to Kirkwall. This would not be acceptable to the time conscious traveller today but it was still a novelty to us to land and take off and we were still comparing the short flights to the long sea journeys.

The flight from Westray to Papa Westray entered the Guiness book of records as being the world's shortest scheduled flight at 2 minutes . This was the time shown on the timetable but as we watched from our hut the flight time with a tail wind was less than a minute on a good day. A rush of camera teams would turn up to film the journey for record breaker programmes all over the world.

The Flying Doctor

In 1965 Westray had gained a new General Practitioner, Dr Mears, who came to Westray from Kent. He had recently gained his private pilot's licence and asked me if it would be possible to join a flying club. I explained that the club in Orkney had ceased and unless it could be revived the prospect of flying looked bleak. With the start of the scheduled air service and access to the grass strip Dr. Mears bought an Auster aircraft which he flew and kept in the open air on Westray unless gales were forecast when he was forced to keep it in Kirkwall.

One day when I discovered him tying the aircraft to a fencing strainer in our field I told him that if the wind increased further the plane would soon be flying like a kite. I then told him that if he bought some concrete blocks I would build him a hangar in the corner of the field near to the terminal which served the Loganair air service. This was finally completed and the Doctor's plane was safe.

The Doctor then tried to get a number of us locals interested in flying and we started doing evening classes in the Westray School with the head teacher Norman Cooper, as lecturer. He had served as a navigator in the RAF during the War.

In the meantime I had written to the flying school at Scone about flying training there and they would have been very pleased to have me but I couldn't get long enough off the farm at that time.

However another RAF pilot, Bob Learmonth, who had flown Spitfires during the War agreed to re-sit his instructors rating as well as examiners rating and the Orkney Flying Club was restarted in Kirkwall using a French Rally Club aircraft for training. Bob became a travelling instructor and agreed to come out once a week to Westray to train us. Because of the weather he didn't manage out every week but came pretty regularly starting on 15th of August 1972. At first there were three of us in the cohort, Tommy Logie, Tommy Pottinger and myself. After a short time Tommy Logie withdrew leaving just the two of us.

One day during circuits and bumps one of the trainees remarked to the instructor on the number of seals sprawled out on the rocks below and was warned to pay attention or they would both end up among them.

We had some hilarity at our evening classes too and one night during the theory of flight we were discussing thrust, lift, drag and so on. One candidate said that that was all very well but in theory a bumble bee shouldn't be able to fly but he just got up and did it! It was all very well for them.

Going Solo

Bob the instructor said that we would have to go to Kirkwall Airport to do our first solo flight there as well as our General Flying Test exam but actually we did it all on the airstrip at Westray. Tommy Pottinger went solo after ten hours' instruction, so I asked Bob to tell me what I was doing wrong. He said I was holding off rather high and then over correcting. When I had made several better landings he said that if I did as well as that next week I'd go solo. I warned my wife that if she saw the plane pass the window with only one person in it to come down to the airstrip and record it on film. When I had done three reasonable landings Bob started taking off his harness as if to leave.
"Where are you going , Bob ?" I asked.

"There's no point me sitting here when you can fly and land like that," he said as he left me alone in the plane. I took off with some trepidation and as the wheels left the ground I knew it was all up to me. When I landed it possibly wasn't a classic landing but it was good enough and I have it captured on cine film. No more actual flying that day although I spent the afternoon floating on air - I had gone solo in twelve hours.

At about forty hours of training an instructor from Inverness came up with a Cessna 150 to teach us spins as the Rallye could not be spun. I must admit I did not enjoy spinning the aircraft and was glad when it was over. The appearance of a small plane hurtling wing over tail towards the ground is more impressive from the ground than the cockpit or so my daughter Margaret told me later. This was to explain why the shock of seeing me stall into the spin had caused her to fall into the ditch as she cycled home. It was a pity I hadn't been able to warn her in advance because for several anxious moments she thought she had seen the last of her father.

There was a moment before my first spin when the visiting instructor suggested I take a piece of chewing gum to help my ears adjust to the spin. I heard Bob say under his breath that I would regret it but it was another new experience for me and I never thought about the repercussions for denture wearers. At that time non-stick chewing gum hadn't been invented and I found myself spinning through the air with my teeth firmly clenched together with gum. In between my first and second circuit I drove up to the house and applied the pot scrubber to my false teeth before I could use them again. I haven't chewed gum since.

Soon the great day arrived for my G.F.T. which I passed successfully with Bob Learmonth as CAA authorised examiner. I must say Bob was a first class instructor who would not let you off with any nonsense - when he was finished with you I reckon you were safe.

From my log book I see my first flight as pilot in command was 30th August 1973 and that my first passenger was my father-in-law. He said to me as I strapped him into his harness that I was not to worry if anything should happen to us as he was well over 80 anyway!

Getting Wings

I borrowed the club plane from Kirkwall for some time and kept it in the Doctor's hangar overnight. Soon after this the Doctor and I bought another Rallye Club Aircraft from Dundee and we shared it with out ever having any argument over its use. The registration was G AXHI, or Hotel India as we called it, and I found it really useful for attending Council meetings in Kirkwall allowing me to get home overnight. Quite often a friend would fly into Kirkwall with me to do a day's shopping until I was ready for home again, or one of our daughters would come home from the school hostel for a weekend and return again for school on Monday morning which was not otherwise possible.

It was during the time we had that plane that there was a sugar shortage on the Island just at the time the rhubarb was in full spate and the strawberries were needing to get picked. Every person who flew with me appeared to have collected several heavy bags of sugar to be stowed in the back seat. I didn't mind as none was spilt but then again I wouldn't have minded if someone had left a pot of strawberry jam either. It was not unheard of to have jam jars in the plane, but let me explain:

When an engine check was needed I had to fly to Perth to have this done taking roughly two and a half hours each way. This is quite a journey in winter when the aircraft has no toilet facilities. Bob and I were both similarly afflicted one time and were forced to come down in a field. Thankfully there were two bushes nearby, but on future trips I began to carry empty jam jars - with lids.

Dr. Mears had left Westray by this time and I bought out his share in the plane by agreement. After a while I was advised to replace Hotel India and so began looking for an aircraft to replace her. The second plane was Golf - Bravo Delta Echo Delta, also a Rallye, and as it happened I bought her from a farmer in the South of England.

The third and final plane was Golf- Alpha X-ray Whisky Hotel which I was able to get through a contact with Dr. Mears, who was by then working as a doctor at Heathrow.

He sent particulars of the plane from Bigging Hill and when I decided to buy it he was very keen to fly it home for me. He took another friend from the CAA with him who was also a Doctor, by the name of Ian Dalziel. They enjoyed the long flight up and there was great excitement when they arrived at Kirkwall. Dr. Mears had been a GP in Kirkwall after he left Westray and received a welcome in his own right. There were photos of the plane in the paper that week and even my wife agreed to come into Kirkwall to meet the plane.

The two Doctors flew it out to Westray with my wife and I followed them out in G-BAOT, the plane belonging to the club. I took Duncan Peace and Freddy Croy with me so that they could return the club plane that evening. As soon as I arrived in Westray we saw the plane at the hangar and I was keen to test drive the plane and so Duncan, Freddy and I set off for a flip. The new plane being a super trainer handled a bit differently and as we were coming round to land I asked the men if they thought I should increase the speed of approach accordingly. We all thought this would be the thing to do but came in at such a rate we were well up the runway and still doing a great speed. On asking later we discovered the approach speed was exactly the same , 60 m.p.h., but at least it had been a consensus decision. I really enjoyed flying that plane and given the right conditions she would even take off in 100 yards.

There was an amusing incident involving the club plane shortly afterwards when it had its photo taken by a holiday maker in Kirkwall. She sent the snap to the British Association of Occupational Therapists because of the BAOT registration on its side. The Association published it in the professional journal and were quite surprised to be informed that the pilot had 2 occupational therapists as daughters.

Calf over Calf Sound

I hadn't been flying very long when I happened to attend a displenish sale, on Shapinsay and bought a cow in calf. I couldn't get her brought to Westray until early the next week because the boats didn't suit. She spent the weekend in Kirkwall and then calved on board the steamer. When the crewmen put them ashore in Westray they put the wrong calf with her - she kicked it all over the place and refused to accept it. I phoned around until I discovered mine was by now in Stronsay.

I spoke to the farmer there and agreed the best thing would be if I flew over to collect it and reunite it with its mother. I asked the farmer to bring it to the airstrip and to ensure it was restrained securely and put into a sack. He agreed to this and so I flew to fetch it. Flying back to Westray I nearly half way home and was flying over Calfsound,

OK A QUICK SHOT
BUT NO BARNSTORMING!

appropriately enough, when the calf's head went past my shoulder trying to get in the front beside me! If he had managed it I'm sure it would have been the end of both of us but thankfully I was able to grab his halter with my right hand while flying with my left. I turned him on to his back and there he stayed upended between the seats with his legs kicking in the air. The plane was shaking about a bit but we made it safely to Westray. After that whenever I carried calves aboard I made sure they also wore a seat belt!

Many years later when I retired to Kirkwall I was needed to fly another calf back to the farm for my son-in-law. I was preparing to circle the farm and announce my arrival when this calf had the same idea. I landed double quick and phoned instead. When we lifted this calf out of the plane we discovered he had wet the seat in his excitement. The calf was always known by the name "Julie" afterwards. Julie was at the time my youngest grand daughter on the farm but she wouldn't let me tell you why the calf was named after her!

The Rosemary

One evening on my way into Church the coastguard sent a man to ask me to come and search for a missing fishing boat called the Rosemary which had broken down, they reckoned, ten miles off the Brough of Birsay.

As Auxiliary in Charge of the coastguards on Westray they often asked me if I would mind using my plane to search. This particular Sunday evening I told them I didn't have a lot of fuel aboard but I would have a go and try to find the casualty. Taking another auxiliary coastguard with me I flew 10 miles west of the Brough and saw nothing. The coastguard station then sent me down the coast of Hoy but again we drew a blank. The third course they set me was to steer out west from Hoy Sound but shortly afterwards they changed this a few degrees and asked me to fly 10 miles out on the new course.

Just as I said to my companion that we were 10 miles out he spotted a vessel below us with the binoculars. I went down to investigate and discovered it was indeed the Rosemary with the net stuck in the propeller. The coastguard then asked me to fly into Hoy Sound to meet the lifeboat and direct her to find the casualty before it became dark. Unfortunately I was so low on fuel that I could not take her to the vessel, but I was able to direct them on to the course that would take them to the fishing boat.

The lifeboat found her and was able to tow her into Stromness harbour before midnight. Hurrying home we encountered a severe hail storm over Rousay but thankfully it was clear over Westray and we got home safely. There is always a great satisfaction in being able to help in the life of our community and we all enjoy a happy ending.

Heavy Weights

With the four seater aircraft I could take three passengers but the combined weight in the back could not exceed 20 stones. This was usually two ten stone daughters with no shopping or a ten stone daughter, her youngest sister and the shortfall in sugar or whatever provision was needed from the shops in town. I grew accustomed to asking ladies their weight before take off but the replies varied according to vanity, honesty and memory. In any event they usually estimated more accurately than men.

One day I was planning a trip up to Shetland with my great friend Alister Hislop who was no light weight. He wanted to meet some friends at Sumburgh while he was in Orkney on holiday from the Isle of Man. As a retired lighthouse keeper he was looking forward to the trip. A local fisherman had broken the prop shaft of his boat and could not get it removed from the coupling. Would I, he wondered, be able to take it with me to Lerwick?
"What weight is it Geordie?" I asked . I was told it was no where near a hundred weight so agreed we would take it with us. It was in fact well over.

We put it in the back seat and ran the runway but could not get airborne. Trying one more time on the downhill runway we did eventually get off the ground but had a very poor climb rate. We took our time getting to Shetland but at least we had gained enough height by the time we got there.

On my first flight to Shetland on 8th May 1975 I had made the journey in 50 minutes and returned in 40 minutes . This time it was no different and the return journey was made with lighter hearts and no prop shaft.

I used to assist the local MP, the late Jo Grimond, in his electioneering campaigns and was pleased to fly him around the smaller islands. One day he rang and asked if I was willing to take him to North Ronaldsay and then on to Papa Westray. I didn't realise that his son Magnus was coming too and that he was the same build as his father and as tall already. The extra weight was a bit of a gamble but I didn't like to go back on an agreement. Leaving Kirkwall was no problem but at North Ronaldsay it was touch and go as we tried to take off. We ran the whole length of the longest grass runway and only just cleared the stone dyke at the end. The six o'clock news could have been very different that night in the build up to the election - a far cry from shaking hands and kissing babies!

A similar situation arose with a musician who had come to Orkney to perform some concerts on the islands but who found himself on Stronsay rather than Papa Westray with only a few hours to go until Curtain up. He begged me to come for him and said he had very few instruments with him. Well, the back seat was full of amplifiers and speakers and what ever was in the flight case was sticking out above our heads. I was unable to take off with the cargo and realised too late I should have insisted he weighed it first. The plane was just fit to do wee hops and was never going to leave Stronsay with all that hardware aboard. I insisted he leave one speaker or else give up the idea of the concert. When he saw the size of his venue on Papay I'm sure one was more than enough. I had also learned not to accept guestimates of weight.

Gusty Grimsetter

I found it difficult in the early days to judge the weather conditions from the forecast and was often on the phone to the Met. Office at Kirkwall airport several times a day before a trip. Even knowing the wind speed didn't always help because of the frequent and unexpected weather changes around Orkney's coast.

On one occasion I remember flying in to a meeting of Directors of Orkney Island Shipping Company. The wind was quite fresh, probably about 45 knots, but a lot more in the gusts. I got into Kirkwall's Grimsetter Airport fine and landed with no difficulties. It was at that moment that one of my brakes stopped working and I was unable to get the plane turned through the wind. When the Tower saw my dilemma they radioed to me to hold on and they would send a safety crew. But, being me, I thought I'd have just one more attempt and tried again but as the plane swung the wind caught her and she dipped her wing on the runway and I had to stop the engine.

Later the Loganair engineer had a look at the plane and phoned me at the meeting to say I had slightly damaged the propeller and it would need to be sent south to be reconditioned which curtailed my flying for quite a while. But how else do you gain experience as to how much you can handle? After that I drew the line at flying in wind speeds of over 40 knots, I thought it was getting strong enough then for a single engine aircraft.

On another occasion the plane was due to go away to Scone Aerodrome near Perth for an overhaul. I flew into Kirkwall to collect Bob Learmonth my former instructor who was going to fly south with me.
"The weather's no great doon sooth," I told Bob "in fact Perth themselves say it's so poor we might no get landed." The cloud base was very low and I was apprehensive.
"Och," said Bob, "Let's have a look." If Bob said that you knew that you could look or, should I say, look out.

Anyway we got as far as Inverness and things weren't looking too bad until we tried to fly inland and there was no way we could do it as the cloud base was right down on the hills. We turned back for the coast and Bob flew her out past Lossiemouth, Fraserbugh and Peterhead before passing Aberdeen - I remember commenting to Bob that either we were too low or the High Risers in Aberdeen were getting higher.

By the time we were as far South as Dundee the sky was as black as night and it really was atrocious weather, so bad that I asked Bob if we shouldn't just land at Dundee while we were winning. Bob was as calm as ever and said we were so close to our target now that we should just have a better look. Bob had been in touch with Leuchars radar who had lost our echo as we passed so low over the river Tay. Bob delighted in winding me up and as we approached the Tay Bridge he asked me to check that the lights on the Bridge were not at red or else we might have to wait to cross!

Not long after this we located Scone and were both filled with a sense of achievement as we landed safely. The Chief Engineer came out to meet us - he was an old friend of Bob's and conveyed some surprise that we had attempted the journey that day. He said "We didn't expect you in this weather, boys. Even the birds out here aren't flying today!"
"It's wonderful what you can do if you try," replied Bob, "but if you don't try it's no wonder at all that you get nowhere." I really would not have attempted that trip without Bob's determination, but as he said it was wonderful.

As I gained experience I attempted this journey on my own. There was a problem flying to Scone on one occasion when my radio set began to malfunction. I couldn't get Wick as I flew down and realised I had a problem. I was able to make contact on Scottish and spoke to them a few times, but drawing nearer to Perth I could not raise them on the approach. I switched back to Scottish and asked if they would phone Scone and ask permission for me to land, but they took no action! I realised that I could not stay up there all day so watched for my chance and landed. I wasn't long down when the tannoy sounded into the hangar asking the pilot of the Rallye to make his way to the Control Tower - I was going to be on the carpet!

The controller wanted to know why I hadn't given them an Estimated Time of Arrival but I assured him I had phoned my ETA down from Orkney earlier in the morning. This was confirmed and soon we were both cooling down.

Flying South for the Winter

In October 1974, during the schools October break, we decided to go on holiday with the plane. I set off from Westray with my wife, Nan, and my youngest daughter Jacqueline who was 5 years old and having her first holiday from school. Her mother had been teaching at the local school on Westray so both were enjoying the end of term feeling. Our second youngest, Fiona was at the school hostel on the mainland and had waited there rather than taking the early morning boat home. She had in fact been up since 6a.m. as heavy rain during the night had found its way in through her dormitory roof and several ceiling tiles came crashing down as they slept.

The morning was dry and clear however and we flew in to Kirkwall and fuelled up. The amount of luggage we needed meant Fiona had to follow us on Loganair's scheduled flight and we all met up at Wick airport where we hired a car. We drove off South after hangaring the plane and we had a really good week. The contrast in scenery is especially obvious in the Autumn with the trees changing into so many different colours. To Orcadians who spend so much time with no trees in view this is quite a novelty.

As we returned to Wick on the Monday it became increasingly apparent that we would not be flying home that day under our own steam. There was a strong northerly wind blowing and I reckoned it was too strong for me to fly over the Pentland Firth. The girls (including Nan!) booked on to the Loganair flight and set off for home, not so much deserting me but all three had to be back at school on the Tuesday morning.

I spent a night with a bus driver who lived close to the airport and on the Tuesday morning the wind had moderated quite a bit and I left Wick, landing at Kirkwall for fuel before taking off again for Westray. However just after take off the Coastguard at Kirkwall came on the Radio to ask me to have a look for a dinghy as I passed the Green Holms as they had reports of a missing creel boat and did not know what had happened to the man aboard.

I made a low pass over the Green Holm and saw the dinghy ashore on the rocks at the north shore of the holm with the oars over the side. Having seen this I made a search over the small island but could not see him anywhere. Apparently as I passed he was climbing up to the top of the island to wave for help but I missed him.

The coastguards then got Jim Stout from Stronsay to go across with his boat, but being deep draughted he couldn't get a landing. As it happened that day there was a helicopter up on lighthouse duty and having heard about the situation he offered to go over and pick him up which he did. He helped him also to get his dinghy safely up off the beach before taking the man home to Eday. I was very relieved to get on home to Westray with everyone safe. After a week's holiday away from it all I felt I was already "back in harness" before I touched down on the airfield.

There was often a feeling on the farm that I was not completely focused on the more general farmwork, preferring to be on the plane or the boat. On one occasion Marcus Hewison who worked on the farm was exasperated with how long I had been away and came into the house looking for me as soon as I got back. He was so anxious to

get me out into the byre that he hurried in to the kitchen and exhorted me to hurry.
"Get thee peedie broon sheun aff and the beuts on!" This meaning loosely that my brown city shoes were not appropriate and that rubber boots were more the thing. It became an in joke in the family and the number of changes of clothing in a typical day - between farming, council meetings, coastguard duties, army cadet parades or yachting - had ensured that I was already a quick change artiste before I became a pilot.

The flying vet

Quite often when we had a sick animal on the farm I would fly over to Sanday where there was a resident vet, Bill Carstairs. I don't really think he enjoyed flying very much but he always agreed to come and do whatever was required. On one occasion I was flying him home to Sanday and as we were coming towards the cliffs known as the Red Heads of Eday he suddenly asked me what would happen if the engine cut out.

Trying to reassure him with a scientific answer I told him that the aircraft would glide for up to thirteen times its height before we had to come down. He stared straight ahead at the Red Heads towering in front of us and I realised he had not found comfort in my answer. Perhaps I should just have told him I had a life raft aboard.

A DeHavilland Dragon lands at Westray. (This one was flown by Capt. Fresson). Photo: Ernest Marwick Collection

Bob Learmonth, my instructor, congratulating me on completion of Long Cross Country. Dr Mears standing by.

Hotel India with hangar I built for Dr Mears. Westray Terminal building is behind.

Hotel India taking off at Westray.

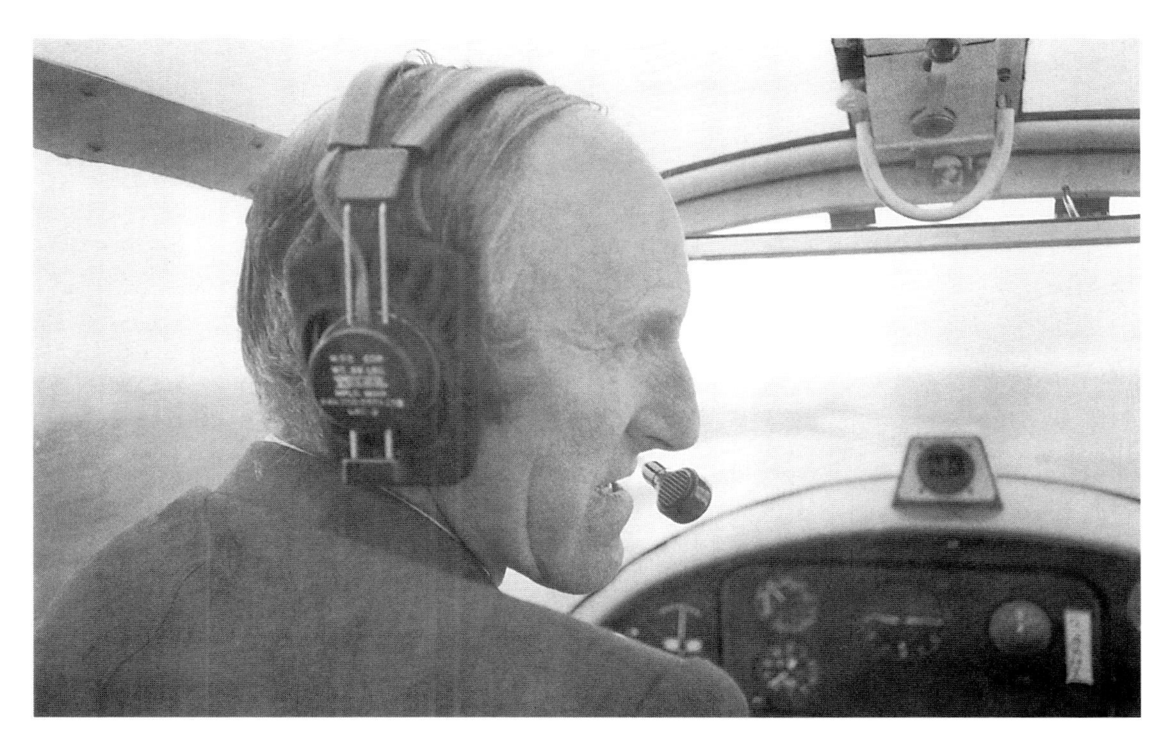

In flight in Hotel India over the sea en route to Westray.

My brother Jimmic and I enjoyed flying when he and Chrisabel visited Westray.

My second plane Echo Delta.

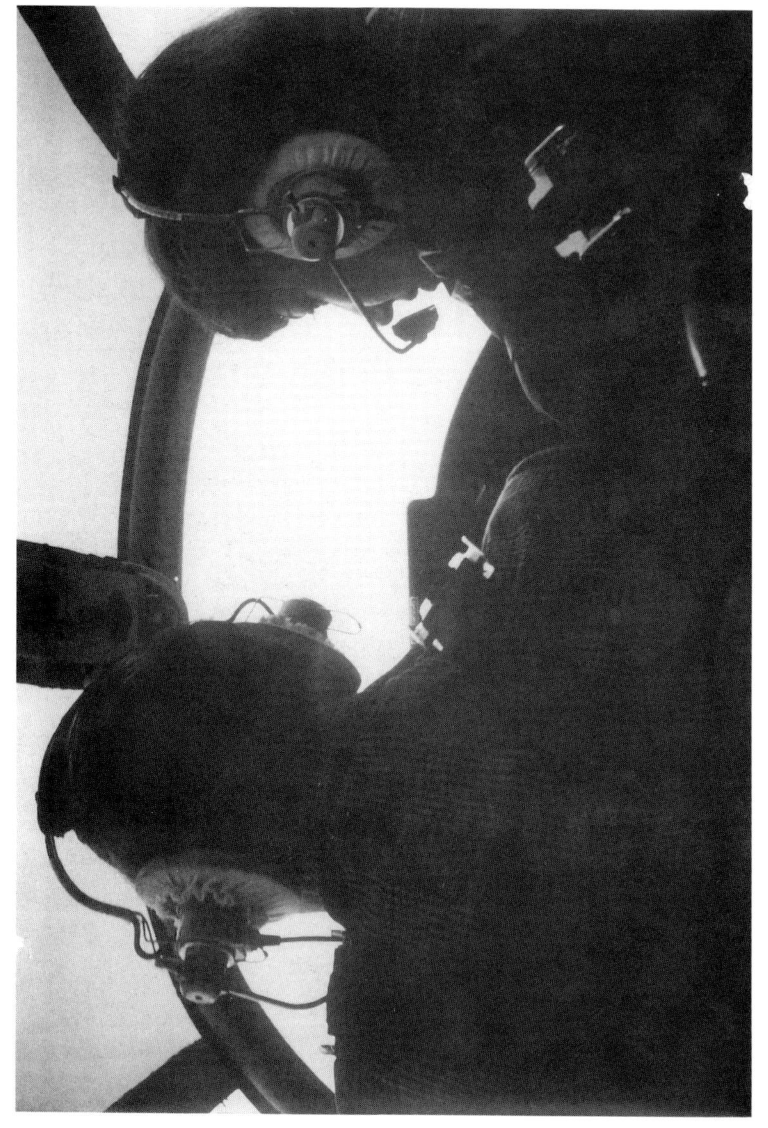

A spare headset makes conversation easier – talking here with Alaine McQueen.

Dr Mears and Dr Dalziel who brought my final plane up to Kirkwall.

Winning the Concourse D'elegance at Kirkwall Air Show. Other prizes were given to David Stephen and Erling Flett.

My wife and I and Geordie Rendall who was over 80 and still loved to fly.

Duncan Peace and I fly the club plane back to Kirkwall.

Panel on Whisky Hotel.

Three of my Westray grandchildren who moved back from Northern Ireland to Skaill. Ruth, Stuart and Julie Hagan.

My wife Nan, after a day of shopping in town.

Whisky Hotel, with Fred Croy taking her to Inverness after my last flight in her, Kirkwall Airport buildings behind.

Another time I remember coming back, after leaving him on Sanday, to Westray. I was at the Red Heads again when the fog closed in all around me. I had to drop down to zero feet and fly over the sea until I found land at the South end of Westray. I was able to find the road and follow it all the way up the Island until I got as far North as the Old Kirk . I had to climb quite a bit to clear the Kirk which stands on the top of the brae. I climbed so high I lost all sight of the ground and I couldn't see a thing. When I judged that I had got over the top I eased her carefully down until I could find the road again. I took a turn as soon as I got my bearings and headed for my brother-in-law's farm at Tuquoy. I knew he had a field large enough to land in. I was very relieved to touch down safely and headed for the house.

As I walked in Helen, my sister-in-law, was at the window wondering to herself what sort of idiot this was flying low. "The sort that needs a cup of coffee, quick please, to steady his nerves," I said. I had never been so glad to visit them and I stayed until the evening before it was clear enough to take off and take the plane home.

About this time we had experienced some problems with calves on the farm and several had died with no explanation. The vets couldn't find what was the matter, so Bill Carstairs said that as soon as the next one died, I was to fly it immediately to Castletown near Thurso where he would arrange a vet to meet me and take the animal to the vet lab in the town.

As it happened the next death was on a Sunday morning. I got some help and put the dead calf into the plane and set off for the airfield at Castletown. Just over Scapa Flow I ran into a snow storm and realising how thick it was I decided it was madness to carry on where I could see nothing. I put back to Kirkwall and rang Sanday to tell Bill that I'd have to postpone it on account of the weather conditions. Bill was very concerned at the prospect of his colleague waiting at the airstrip in Caithness and told me so. "Surely Bill," I said "my life is worth more to you than a dead calf?"

I have no idea what thoughts he had on the matter but I decided that my fate was of more importance than the calf whose fate was already sealed. I set off for home as soon as the weather improved. Next day we put the calf by Loganair to Wick and thence on to the lab on Monday morning. After a long time they discovered it was a rare virus which had only been in the country for a couple of years and no one knows how it appeared in Orkney. I feel obliged to point out that it was not BSE!

Shortly after this my wife had two foreign gentlemen call at the farm selling goods from a van. As we were the most northerly house on the Island she offered them a cup of tea and some of her newly made thin scones with strawberry jam from the garden. This was so much appreciated that she was presented with a bead from a 'most Holy place in Bangladesh'. It appeared to be one from a long string of similar beads but with great ceremony she was told that if she kept it safe 'no more cows get sick and old Granddad live to be very old man'. He was already approaching ninety but as it happened the virus among calves died out of the herd and, whatever the reason, we were all very relieved.

It was not only the vet who found himself summoned and flown off to operate at short notice. One day Dr Mears had invited the county Surgeon out to see a patient on Westray who was not fit to travel in to the hospital on the mainland. He had asked me to take him in to Town afterwards saying it was crucial he get back to the hospital as he had another person to see that day. We set off just after 5 o'clock but no sooner had we taken off than fog appeared. Even the thought of a surgeon aboard didn't encourage me to take a risk and we went back to Westray for a bit until it cleared. However we got into town a short while later and home again safely.

Air Shows

In August 1980 there was an air show planned for Kirkwall Airport. The flying club in Kirkwall was going well and there was a lot of interest. The promise of seeing the Red Arrows was a great attraction and many local folk travelled to Grimsetter to watch, including those who would normally have found chores to prevent their better half from attending.

All the Orkney planes were invited to come in for a fly past along the main runway and although we were very tame entertainment after the antics of the Red Arrows there was a lot of local interest, and we all enjoyed taking part. Less enjoyable was the spot landing competition which involves as the name suggests coming in to land on a given spot. What was possible to practice in your own field with only the cows for audience was more challenging with crowds watching, knowing as I did how many pilots were among the onlookers!

I did on one occasion win a competition at the Kirkwall Air Show, but it was for beauty rather than technical merit. In fact it was the plane and not me who attracted the judges attention and won the Concours d'elegance. At this time I was flying my second plane Gulf - Bravo Delta Echo Delta, which had been cleaned for the occasion. I always felt I had a disadvantage the Kirkwall pilots didn't have as leaving Westray's grass strip always involved having to taxi through cow pats. The pilot who won the spot landing at that show was Erling Flett one of the club's youngest members and now a successful commercial pilot.

There was also a big air show at Sumburgh in Shetland and we were all contacted and invited to attend. I was not sure if I would go until the last moment, as I was having problems starting the plane. It took a lot of effort and you could never be sure that it would start at all until we discovered it was a problem with the magneto.

In the end I flew as far as Kirkwall and it started fine, so Fred Croy joined me and he flew my plane up to Sumburgh, where we had a really great day. The folks in Shetland were keen to use my plane for the flight demonstration but-you've guessed it would not start at all and they had to use the Cessna 172. In the afternoon the fog came down and the whole show was grounded. I went up to the Met Office to see what the forecast was and then to the tower to phone my wife and tell her I wouldn't be home that night. We had been invited out to dinner in the evening and I wanted to give advance warning so that things could be cancelled. She told me there was no fog near Westray where it was a lovely clear day, would I try again later?

I went back to the Met Office with this information and they agreed that if I could fly through the fog as far as Fair Isle (which would be pretty dodgy) it should be fine south of there. It was still really thick but I set my VOR for Kirkwall and set off, with the plane starting first time. I got to Fair Isle and the sun broke through, in fact it was so clear towards Orkney that I could see Westray from the air. I was delighted to get home and as always when you face something of which you are uncertain the feeling you get when you are successful is tremendous. The tension of the day and the relief of being home meant I was really hungry and delighted that the food had not been cancelled.

We also had a fly in one weekend on Westray for some other members of the Orkney Flying Club. While not in the air show league it was still great fun to spend the whole weekend either flying or talking about it. We did stop to eat of course and the food stuff was mostly collected on the grass airstrips! That year was a particularly good one for mushrooms and when I called down to visit them in the hut where they were staying the frying pan was going flat out with the ones they had gathered that morning. When they flew back to Kirkwall on the Monday morning they had obviously been out at them again and were flying home with more than happy memories. I must add that these were ordinary, common or garden, mushrooms and in fact I doubt if many of us had heard of magic mushrooms at that time, and so when I said that the boys flew home, I did of course mean in an aircraft.

That was an exceptionally good year for picking mushrooms; some days the field was almost white in parts. Even the Loganair pilots started bringing out poly bags to take home their harvest although what the passengers thought as the pilot jumped out at the end of the runway and started picking mushrooms I never heard. I never saw such a good year since either so whatever freak conditions caused the crop have not recurred.

Friend or Foe in the Westray Firth

There was a ship called the Mercurious which often passed Orkney as she carried her cargo of cement. The Dutch captain was keen on CB radio and always wanted to talk to Orcadians when he was within range. I had often had a copy with him on the CB at home, and speaking to him one night I realised he was passing fairly close. I asked if he would mind if I flew out and took a picture of the ship from the air and he was more than happy about this. I made a low pass as he approached Westray and got some good shots.

I heard a while later that she got into difficulties en route to Stornoway and eventually sank, however that particular captain was no longer serving on her at that time. Although we never met he kept in touch and was always keen for me to visit him in Holland although, so far, that has never transpired! His C.B. handle was Black Tulip.

Another time during a NATO exercise off Orkney I was flying home to Westray one evening when I spotted a submarine going up the Westray Firth. I was unaware of the NATO presence and called the control tower to advise them of what I had seen.

I thought no more about it and it was not until many years later when I was retiring from the local auxiliary coastguard that the matter was cleared up. In making a presentation to me the visiting official told the story in his speech. How my report reached the coastguard I don't know but apparently the fact that I happened to be passing overhead foiled the whole operation and the submarine had been unable to pass Orkney undetected.

There are very few secrets in Orkney and people are used to living under the surveillance of their neighbours. My daughters were always very interested to see their friends houses from the air and would gaze in safety at the back gardens or behind the bike sheds - no need for net curtains in the air.

Surprise surprise!

Something that I always enjoyed was giving Nan a surprise by bringing one of the girls home unexpectedly from school or university. On at least one occasion I watched them eat the fatted pork chop or whatever had been left for my meal. The scheduled flights left no room to surprise local folk, even plain clothes policemen were unable to get to the Island unexpectedly and before ever they landed every untaxed vehicle would be off the road and bald tyres safely in the garage, out of sight.

Just after New Year in 1982 I had a phone call from Mike Bishop, a fellow known only to me as the boyfriend of my daughter's best friend. He had been unable to get up to Papay to join Alison for Christmas and New Year and was supposed to be on holiday in Aberdeen with his father. He was calling from Caithness and told me that he and his dad had driven up North and stayed in a B&B the previous night. On the spur of the moment he decided to ring me and ask if I was willing to fly them over as a surprise to the folks on Papay.

It was a lovely crisp morning with excellent visibility and I knew I could make it to Caithness with no difficulty - I was more than keen to pick him up. The day was so still that even the wind seemed to be observing the public holiday! It was a great trip over to collect them, and I met them as arranged in great secrecy. I had recollections of Mike from my daughter's wedding the summer before, but I had never met his father, a man by the name of 'Bud' Bishop.

Bud was looking very disconcerted at the size of the plane. He had been expecting something much larger and Mike had to reassure him that these craft were as routine as cars to Orkney folk and that Alison and Fiona used to travel between Papay and Westray 'all the time' visiting each other. He got in.

I think he was beginning to relax a bit as we flew out towards Papay. We decided not to touch down in Kirkwall in case word got out and as we approached Papay Mike leaned over and told his father that it would be a grass airstrip when we got there. I think he was in two minds, one being to demand I take him back; the other that I open the canopy and let him jump from a safe distance. The thought of us hurtling along a small grassy field trying to stop before we took down a stone wall was too much for him. He closed his eyes and missed a perfect landing.

Now the story had a happy ending, Mike made amends to Alison for missing the usual hogmanay festivities on Papay and a good time was had by all. Mike accompanied me back to Kirkwall on his way south although his father in seasonal style followed the example of the three wise men and 'went home by another way'.

The Man fae the Ministry

Many local professional people used Loganair for travelling between islands in the course of their employment. It was often a novelty for the first few months before it became as routine as the bus. To newcomers however it was not as predictable and to one young man from the Ministry of Agriculture and Fisheries it was completely alien.

Alisdair had spent a busy summer's day on Sanday driving round various farms in a hired car. At what he thought was the appointed time he drove towards the airfield to watch his plane leaving the ground without him. The cars were leaving the hut and only the two attendants remained to inform him that the plane had left. They thought the pilot was in a hurry because there was fog out to sea.

“How long till the next one?” asked Alisdair, looking at his watch.
“Hid could be long enough, best kens, but no more the day.” The attendant predicted.

Alisdair patiently asked what other options there were for getting home that night to discover that there were no more flights or boats and that the attendants were going home for tea. Before they went they said that Jack Scott sometimes gave the vet a lift and so Alisdair got on the phone thinking I was an air taxi service. The forecast was reasonable and so I decided to go - one of my daughters knew him from Kirkwall and assured me he was very deserving of assistance.

I collected him as agreed but he was obviously not expecting me in my farming clothes. I told him I had come right away to get home before the mist and he was soon more confident that I was a pilot whatever the state of my breeks might suggest. It actually got misty as we approached Kirkwall and I was in a hurry to get him down and turn for home. He was on the point of asking for my charges and wondering if I would take a cheque but it would have taken too long to explain that I was only a Private Pilot. He advised me that he worked for the Ministry, and was wondering if they had an account? I had no intention of sending a bill but nodded as I was hurrying to leave.

I did make it home that night but only just. The fog was appearing in great lumps and I had to island hop all the way home, flying over each island and only heading for the next as a clear patch appeared. He obviously never fully figured out that I was not in business and some weeks later he met my daughter on the St Ola as she sailed off to College in Edinburgh.

"Your Father collected me when I was stranded on Sanday," he said, "and I never got sent a bill". Fiona saw a great opportunity and seized it.

"Well he doesn't fly commercially, Alisdair, but if you are feeling bad you can help me carry this muckle Orkney chair off when we dock at Scrabster." This was the right thing to ask, the fellow was also going on to Edinburgh with an empty car. He repaid the obligation in full, driving her and all her luggage to her flat and stopped to treat her to an excellent meal of venison en route. Cast your bread on the waters?

The next time he visited Sanday he missed the plane again, this time he was late because he had stopped to help a farmer with a difficult lambing. After the confusion of the last trip he decided to cut his losses and headed for the island's only hotel where he enjoyed dinner, bed and breakfast for the princely sum of £2.10!

Nuts, bolts and loud bangs

One day when I was on my way into Kirkwall, the Tower called saying that Loganair wondered if I could possibly cross to Wick as they urgently needed to get their engineer over there where the Islander had broken down. I said I was due for a council meeting but I would try as long as I could get right back. I took the engineer to Wick and when I flew in to Kirkwall the Loganair staff had a taxi ready to whisk me off to my meeting, which was very kind of them.

I always felt obliged to the engineers for their on the spot diagnosis and help. After years of island life I was used to fixing most things on the farm myself. In aviation it really is a different affair and there are not many things on a plane that you can fix with binder twine. It is also difficult to get out and push at two thousand feet and for that reason if no other the standards for maintenance need to be high.

Once Echo Delta, my second plane, had to go to Edinburgh for overhaul and I was having trouble getting her to start as there was something wrong with the starting pulley. They had a look and put on a replacement starter before sending her back to me. When she arrived at Kirkwall she would start but only fired and then cut out.

Loganair's engineer had a look and set the mixture a bit richer and then she ran fine. He said I could take the aircraft home but advised me not to fly over water more than I could help. I flew round the Mainland on the way home then crossed from island to island until I arrived safely at Westray.

When I flew back to Kirkwall some time later I heard a loud bang just over Shapinsay and watched the rev counter drop to zero before coming back to its position again. I knew something had gone wrong but couldn't find a fault and all the other meters were registering fine. We got to Kirkwall fine where I got out but I told Tommy Pottinger not to fly on to Stronsay as he planned until the engineer checked it.

He drew off some oil and discovered that there were bits of metal in it and so we could not fly it again. I had great trouble getting compensation for it and had to get the engine replaced. After we found a suitable one the company agreed to come up and fit it - but they wanted £1,000 to do so! After this she flew very well with no more loud bangs.

Retirement

My wife actually retired before me and took early retirement from her teaching job on Westray. She felt I should be slowing down too and would have been very happy if I had decided to hang up my wings before we left the farm. I was able to help her however when she was summoned by the Education Department to go and do relief teaching in the one teacher school on the island of Papa Westray next to us.

Although she enjoyed her retirement she was very tempted to accept supply teaching on Papay because she enjoyed the atmosphere of single teacher schools. She had been the Headmistress of Skelwick School on Westray when it closed and had 7 classes to supervise.

She was contracted for six weeks in total and travelled over to the island on a small boat run by a local lad. One morning it was too windy for him to attempt to cross so she went aboard the Orcadia which was scheduled for Papay. The steward told her however that he doubted if they could land her there and might have to take her as far as Kirkwall. She returned home worrying, by now, about her pupils.

I told her that as the wind was westerly I might be able to fly over and deposit her in the field if she could walk up to the airstrip. It was a test of her professionalism that she agreed to try it. I flew her over and landed into the wind in a very short distance.

I opened the canopy and got her to jump out, handing down her books and bags to her as she stood, knee deep in grass and other agricultural matter with the wind tearing at her headscarf. She watched me carry on, taking off into the wind, and never took her eyes off me until she saw the plane in the distance being safely flown back into the hangar on Westray. She then stepped (carefully) through the wet grass up to the airport ready to do a day's work.

Needless to say I had to make my own tea that night as the same trick wouldn't work in reverse and the wind continued westerly all day. Nan stayed with a friend and that was the only time I ever flew her to work.

She never wanted to learn to fly herself but almost had a go one day when I had to summon her to help me start the plane. The usual problem of the plane firing but failing to continue running was about to make me late one day for an appointment on another island. After several goes I realised I couldn't start it without help. I hurried to the house and fetched Nan.

"Wait until you hear it fire, push in the throttle, and then pull it back," I shouted as I swung the propeller with Nan in the cockpit. She heard most of this but not the bit about pulling it back. The engine opened and revved up, I hurried round the wing to pull it back but thankfully I'd had the presence of mind to put 6" blocks in front of the wheels. It would have put more than my gas at a peep if Nan had flown solo without any instruction!

After I retired from the farm we moved into Kirkwall and I didn't have the same need or excuse to fly. What had been a necessity for so many years became a luxury and I didn't fly so much. One or two trips did give me pleasure; especially when our minister's mother let it be known that she would love to fly. She was almost 90 at this time and was blind.

We got her into the plane and headed off for the West Mainland. She was having a great time and her son Francis Gordon described everything to her as we passed it. She was really an amazing lady and was game to the heels.

I decided to sell the aircraft with much heart searching. The cost of hangarage was an extra cost I hadn't had to worry about before, and as my wife said, it was not a pensioner's hobby. The landing charge also went up from 50p to £12.00 meaning the flight from Westray was cheaper than the landing which seemed ridiculous.

I advertised the plane and a chap from the Midlands came to have a look at her and was very pleased. He was on the point of buying her so I suggested we take a test flight out to Westray. It was then he noticed a spot of corrosion on the rail of the canopy. This put him off so we flew the plane down to Inverness to let the engineer have a look at her. His advice was to scrap her and sell her for spare parts which did not appeal to me at all.

After a bit a pilot from Devon contacted me as his engine had expired and he wanted mine to replace it. This was fine until I stated a price and eventually the two bidders were told they could decide for themselves who wanted it. After I went to bed that night the fellow from the Midlands phoned and agreed to buy it. He came to Inverness to collect it and all went well until he flew into airspace near Newcastle where he should not have been.

I knew nothing of his adventures until I received a letter of official complaint telling me to phone and explain my actions. I told them I hadn't been near Newcastle and gave them the name and number of the new registered owner. In response a second letter arrived informing me they would agree to take no action this time but that I should never let it happen again! I did not want to end my flying days with a black mark on my record so I wrote to let them know that I had never made a navigational error and was not prepared to accept this one.

Before I sold her I went out for a last flight to Westray and filmed the house and the old haunts from the air, before getting my son-in-law to film the plane from the ground. Flying back in from Westray was the last flight I made, and I have never been in charge of a light aircraft since.

My licence expired and I missed flying terribly. I suppose I should feel grateful that I managed to fly at all especially starting in my fifties and that in sixteen years I flew mostly over sea and never had any engine problems, apart from icing, to cope with. In 16 years I amassed 1027 hours in the air and did 3262 landings! Because of the short flight times entailed between Westray and Kirkwall this stretches into 7 log books which are full of 20 minute entries. Had I been crossing the Atlantic rather than the Westray Firth I wouldn't have half filled my first log yet!